Hi, my name is Little Mo;
I work with Diffo Dynamo.

We pack boxes, big and small,
then stack them up against the wall.

While I keep up my steady pace,
Diffo wants to race, race, race.

Sometimes Diffo's nearly done, when I have only just begun!

I tell him he should take more care,
or he'll have boxes everywhere.

Poor Diffo! He won't look so smug when the boss pulls out his plug.